Friedrich Teichmann

Network Focused Integration of Geospatial Data in Avionics

Friedrich Teichmann

Network Focused Integration of Geospatial Data in Avionics

Concepts and Perspectives 2007

Reihe Realwissenschaften

Impressum / Imprint

Bibliografische Information der Deutschen Nationalbibliothek: Die Deutsche Nationalbibliothek verzeichnet diese Publikation in der Deutschen Nationalbibliografie; detaillierte bibliografische Daten sind im Internet über http://dnb.d-nb.de abrufbar.

Alle in diesem Buch genannten Marken und Produktnamen unterliegen warenzeichen-, marken- oder patentrechtlichem Schutz bzw. sind Warenzeichen oder eingetragene Warenzeichen der jeweiligen Inhaber. Die Wiedergabe von Marken, Produktnamen, Gebrauchsnamen, Handelsnamen, Warenbezeichnungen u.s.w. in diesem Werk berechtigt auch ohne besondere Kennzeichnung nicht zu der Annahme, dass solche Namen im Sinne der Warenzeichen- und Markenschutzgesetzgebung als frei zu betrachten wären und daher von jedermann benutzt werden dürften.

Bibliographic information published by the Deutsche Nationalbibliothek: The Deutsche Nationalbibliothek lists this publication in the Deutsche Nationalbibliografie; detailed bibliographic data are available in the Internet at http://dnb.d-nb.de.

Any brand names and product names mentioned in this book are subject to trademark, brand or patent protection and are trademarks or registered trademarks of their respective holders. The use of brand names, product names, common names, trade names, product descriptions etc. even without a particular marking in this work is in no way to be construed to mean that such names may be regarded as unrestricted in respect of trademark and brand protection legislation and could thus be used by anyone.

Coverbild / Cover image: www.ingimage.com

Verlag / Publisher:
AV Akademikerverlag
ist ein Imprint der / is a trademark of
OmniScriptum GmbH & Co. KG
Bahnhofstraße 28, 66111 Saarbrücken, Deutschland / Germany
Email: info@akademikerverlag.de

Herstellung: siehe letzte Seite /
Printed at: see last page
ISBN: 978-3-639-87039-8

Abstract

For efficient and effective aircraft operations various geospatial data have to be available under predefined conditions. The standard aeronautical in-flight navigation charts, approach charts and aerodrome maps are supplemented by geo-referenced data tables like those indicating location ID, airspace extent, obstacles and navigation points. The geospatial database is further augmented by mission related data like way points, remote sensing data or results of location based services. The integration and proper updates of the geospatial data for aeronautical systems on the ground as well as in the air are critical for optimal aircraft operations. Due to the special requirements of air operations, such as for the various types of air space users, the third dimension (z), the long distances, the required actuality and authenticity and the high level of system integration, the indispensable geospatial data has to be handled uniquely.

The trend in handling geospatial data for avionic users ranges from simple analogue navigation charts over independent analogue geo-systems to the current digital systems which are linked to position instruments like GPS. The future trend might be towards advanced GIS functionality in fully integrated or embedded avionic systems. Embedded avionic geo-information-systems have special hardware constraints, advanced GIS-functionally and should be able to be fully integrated into other avionic systems. This integration would provide valuable inputs like navigation data, time and elevation. Additional links could exist to flight management systems for mission related data transfer and broad databank access.

The highest innovation benefit would result through a network operation centered on the geospatial data. Based on the required data flow and various update rates, three main channels for network centric operations for an embedded avionic geo-system might be relevant. Secure in-flight communication with data link functionality from the aircraft to the appropriate ground operations control unit would enable service oriented architectures with an on-demand specific information management. Secure LAN or wireless LAN at an airbase would permit automated download and update processes of data to-and-from an aircraft and thereby guarantee, that the latest geospatial data is loaded. Open communication, including a routine position report with predetermined procedures, from air-to-air or to the responsible air traffic control unit would greatly improve the situation awareness for all contributing units.

To optimize air space use and mission efficiency while minimizing risk of accident, energy and emphasis must be applied to generate and provide the necessary geospatial data. This data should be integrated into an embedded geo-information-system within the avionic surroundings and be applied for processes in a network which is focused on geospatial data.

Primary Source

Network Focused Integration of Geospatial Data in Avionics; Master's Thesis for Professional MSc. in Telematic Management; Danube University Krems; Submitted by Friedrich Teichmann; June 2007.

Acknowledgements

I wish to thank my advisor, Dr. Häusler, Department of Geosciences, University of Vienna, who supported this research concept completely, was open for new ideas and focused me with well placed questions on the key aspects of this thesis. His input was invaluable.

In addition, I would like to thank my employer, the Austrian Military for its financial support and flexibility during course work and research time. The Air Force Command (Kommando Luftstreitkräfte) and the Ministry of Defense Transformation Department (Management ÖBH2010) played key roles in this project.

The staff at the Danube University of Krems proved to be continually helpful in many areas and showed a remarkable adaptability with regards to a somewhat different, yet composite thesis.

My wife Margaret deserves also to be acknowledged. She not only edited my English, but also provided a stable working environment at home and supported my efforts toward further education. My three daughters enjoyed the concept that their father was going to school and doing homework the same way they were.

Table of Contents

List of Tables

List of Figures

Part I

Introduction to Geospatial Data and Geo-Systems for Aviation

1 Problem Scenario

Geospatial data are among the most crucial information elements for successful aircraft operations. These are applied by the pilot in the air as well as by the ground control center. Despite the fact, that numerous operational aviation elements handle different types of geospatial data, a number of limitations in the use of geospatial data exist. These deficiencies are either a result of misconception of the range and special characteristics of geospatial data or are based on limited system integration and minimal network operation processes.

Geospatial data in aviation cover a broad range from aeronautical charts to tables with radio navigation points to mission specific high resolution remote sensing data. An additional level of complexity in handling these different types of geospatial data arises when their integration into a modern avionic system is desirable or required for mission efficiency. Complexity is further amplified, if the resulting geo-information-system is incorporated into a more modern, innovative and multidimensional network operation.

Although much emphasis has been placed on providing the geospatial data needed, currently neither the pilot using avionic systems nor most ground control centers have integrated or implemented network centered processes. These would provide all discussed geospatial data in a timely manner for aviation operations.

2 Aim and Scope of this Study

The concept of an innovative, integrated geo-information-system with up-to-date geospatial data which provides service oriented applications and on demand information management has to be defined. To best describe a potentially embedded avionic geo-system, one has to link advanced GIS with the preexisting avionic systems. The necessary relations for an upcoming network operation can then be defined.

Geographic data can be generated and/or resourced by a variety of disciplines. These could be pilots, navigation instructors, aeronautical cartographers, GIS-specialist or air traffic controllers. These geographically and sometimes non-geographically schooled individuals apply and interpret the data differently and according to their needs and projects. The general concept of this study is to discuss the integration of relevant geospatial aeronautical data from a geographical viewpoint.

The scope of this study is three fold. The first step is to analyze and evaluate the different clusters of geospatial data for aviation as well as to determine the characteristics of this data. This step has to be supplemented with information about how geospatial data can be made available to the aviation community. This ranges from handheld analogue maps to full IT-system integration. Accordingly, the clearly evident increase in complexity of the resulting geo-information-systems in aviation must be taken into consideration.

The second section of this thesis focuses on some special aspects of geodata used in military aviation. These have been independently published by this author in the International Handbook Military Geography. The focus of this section is on the unique properties of geospatial data for aviation, the diverse aeronautical charts used during flight and a review of the additional geospatial data used for military aviation. Despite the fact, that part II focuses on military aviation, all discussed aspects can be applied in many areas of civil aviation as well.

The final section of this study focuses on system integration and network operations with the geospatial data located at the nucleus. The necessary hardware constraints and advanced GIS functions are evaluated and the data flow for an integrated avionic geo-system is analyzed. Building upon these advancements, the final part of this section is on network operations which then yield an integrated aviation information system. The following three clusters are important for the envisioned network operation: secure in-flight connection between air and ground control, secure (wireless) LAN at the air base and open air-to-air in-flight communications.

This research is focused primarily on aircraft avionics and pilot perspective regarding information management. The focus is not on the ground support systems because additional variables may be involved. Despite the standardizations in aviation, the terminology and the discussion in this study applies largely to Europe.

This study is not intended to evaluate the processes and organizations required to provide the various geospatial data to the aviation community. Furthermore, this study does not handle technical details like protocols, but instead presents interpretations and ideas at the system level.

3 Practical Applications

This study presents new aspects for handling geospatial data in a network operation environment. The benefits of the discussed improvements are for both the pilot in the air and the ground operation centers.

Certain civil air missions, like Search and Rescue operations, require almost identical geospatial data integration, while civil air fleet operations rely very heavily on the emerging network operations (Hughes 2005, 2006). Due to the high level of standardization the envisioned improvements can be applied for civil as well as for military aviation.

The ultimate goal is to implement an embedded avionic geo-information-system with a network centered approach to handling geospatial data. This would facilitate the planning and operating of an aircraft fleet, improve the safety and reaction time in the air and enlarge the economic advantage in aircraft operations.

4 Historical Developments of Geospatial Data in Aviation

Aeronautical charts are the most obvious contributions to the avionic geospatial database of all types of geospatial data. Practically every cartography textbook has basic information about the different aspects of aeronautical or navigational charts (for example see Reents 1957, Wilhelmy 1981, Leibbrand 1984, Hake 1982, Hofmann-Wellenhof 2003). The term chart dates back to the sea faring days when nautical charts were developed. Its primary purpose was navigation. In addition to the previously mentioned cartography textbooks, a large number of books, manuscript and guides are available for navigation for pilots (For examples: Hitchcock 1997, Clarke 1998, Hall 2006, among others). Both aeronautical charts for en-route navigation and approach or terminal charts for the final flight phase are standardized for use in the international aviation community (see International Civil Aviation Organization at URL 1, Federal Aviation Administration at URL 2, Eurocontrol at URL 3 and Jeppesen at URL 4). The standardization does not only cover the layout, content and display of the charts, but also the distribution cycle and system. As seen in Figure 1, the primary and most evident content of the geospatial database are its aeronautical and approach charts.

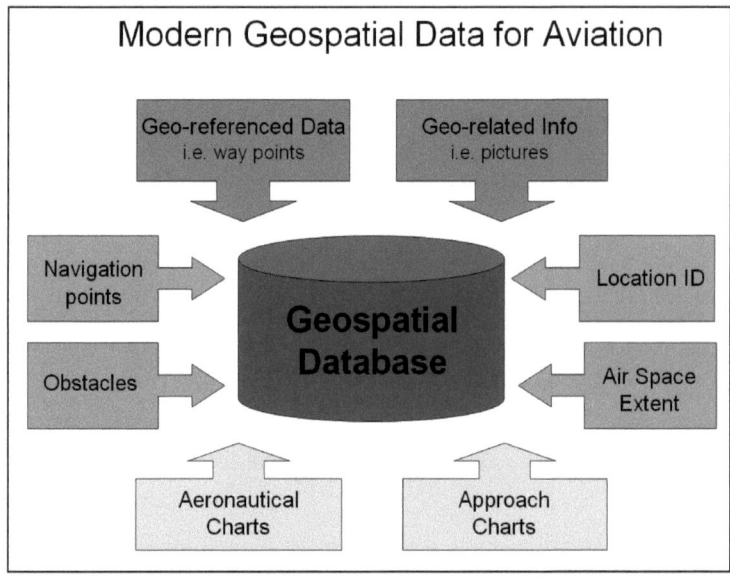

Figure 1:   Contributions   to   Modern   Avionic   Geospatial   Databases (Teichmann 2007)

In addition to the two graphic essentials for navigation there are important aeronautical tables (Gassmann 1989, Kant 1989). These make an equally important contribution to the geospatial database for aviation. All points of interest of the aviation community, which have been geo-referenced, are included. The primary contributions of tabular geospatial data to the database are the location identifiers, the air space delineated by cartographically defined borders, obstacles such as aeronautical hazards and all (radio) navigation points (Figure 1). The principal owners of these tabular data are the national aviation authorities. The navigation points (in connection with air space extent and location ID's) are a key element of the location communication between the air traffic control including the aviation authorities and the pilots in the airplanes. Depending on the flight rule (IFR versus VFR) some tabular geospatial data can be more important than other and this can also vary from user to user. Although the format of the tabular geospatial data is generally standardized, the process of how to implement it, for example the obstacles data base, is variable. This occasionally leads to the effect, that the discussed tabular data is not universally recognized to be geospatial data. Despite the fact, that these data are clearly defined geospatial data, they are mainly applied only in air traffic control processes.

The final two data blocks (Figure 2) contain additional geospatial data which are only marginally interesting to air traffic controllers or aviation authorities. These include geo-referenced mission data such as way points and geo-related information like remote sensing data (Figure 1). These two data blocks can vary tremendously depending on the specific mission to be carried out and might even be the key elements for success. Search and Rescue or security missions without either geo-referenced tactical information or detailed pictures (For example: a helicopter landing site) may not be carried out successfully. These additional geospatial data are not standardized and every organization dependent on them should implement appropriate procedures to provide the necessary data.

The key point is as follows: The aviation community must strive to attain a comprehensive geospatial data package, which ranges from charts to mission way-points and is integrated into a common geospatial database. Appropriate processes have to be used to implement and to update this central geospatial database. This process could even be automated. All implemented aeronautical information systems on ground (air traffic control) or in the air (avionic) must have access to this up-to-date comprehensive geospatial database and must be able to extract the necessary information in a standardized way. Through this, the ultimate goal, which is to minimize risk using integration of high level geo-data, can be realized.

5   Development Trends in Geo-Systems

The concept of a comprehensive geospatial database for aviation is concurrent with the development trend for geo-information-systems in avionics (Figure 2). The simplest form of a geo-informaton-system in a cockpit is composed of analogue charts and tables, which are currently standard equipment for all pilots. During the last quarter of the 20th century, modern flight management systems with autopilots were implemented in avionics (Figure 2). As a supplement to analogue navigation charts, some specialized airplanes were outfitted with pioneering cartographic display elements which were mounted in their cockpits. The technology employed was based on a mechanical coupling of a positioning system such as an inertial navigation system with a high resolution optical output element which uses a prefabricated map on a slide.

The next important development step was the availability of satellite navigation systems based on the American GPS system (Figure 2). This technology (Blanchard 1995; Grewald 2001) in conjunction with rudimentary geographic information systems allowed a digital display with on-time positioning for the first time. Most low cost systems employed in aeronautics are independent digital geo-systems and use the GPS-positioning on a mobile GIS-unit.

Current developments and the focus of this research are geo-information-systems, which are integrated into all other relevant avionic systems and thereby provide numerous information advantages (Bartels 2006, Phillips 2006). Part III of this study will elaborate upon this concept.

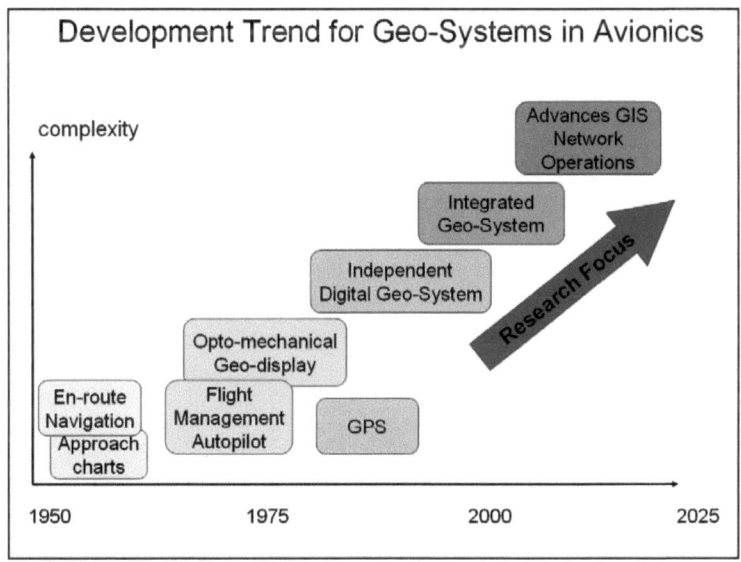

Figure 2:   Development in Geo-Systems in Avionics (Teichmann 2007)

The optimization of avionic information management could be carried out through system integration and, based on the new technology currently available, on modern hardware and software implementation. The future development path will then focus on optimizing the data flow (Gruender 2003). To achieve this, geospatial data will have to be made the central focus of network operations.

6   Definitions

Aviation
- The science of flying and operating aircraft
- Aircraft manufacture, development, and design
Cayne, 1993: New Webster's Dictionary and Thesaurus

Avionics

The application of electronics to aviation and astronautics

Merriam-Webster Dictionary (URL 5)

Database

A logical collection of information that is interrelated and that is managed and stored as a unit, for example in the same computer file. The terms database and data set are often used interchangeably. A GIS database contains information about the location of real-world features and the characteristics of those features.

UN Handbook on Geographic Information Systems Digital Mapping (URL 6).

Geographic Information System = GIS

A collection of computer hardware, software, geographic data, and personnel assembled to capture, store, retrieve, update, manipulate, analyse and display geographically referenced information.

UN Handbook on Geographic Information Systems Digital Mapping (URL 6).

Georeferencing

The process of determining the relationship between page coordinates and real-world coordinates. Georeferencing is necessary after digitizing, for example to convert the page coordinates measured in digitizing units (e.g., centimetres or inches) into the real-world coordinate system that was used to draw the source map.

UN Handbook on Geographic Information Systems Digital Mapping (URL 6).

Geospatial

A term that is sometimes used to describe information of a geographic or spatial nature.

UN Handbook on Geographic Information Systems Digital Mapping (URL 6).

Spatial Data

Information about the location, dimensions and shape of and the relationships among geographic features. In GIS, spatial data are technically classified as points, lines, areas and raster grids.

UN Handbook on Geographic Information Systems Digital Mapping (URL 6).

Geo-system, geo-information-system

Geo-system and geo-information-system are both terms used frequently throughout this thesis. They have been used to emphasize the modern network system structure and can be interpreted to be expanded definitions of the standard GIS.

Part II

International Handbook Military Geography
ISBN 3-901183-50-7

Teichmann, Friedrich

Geodata for Military Aviation

Page 426 - 443

Keywords

Aeronautical chart, aeronautical information, military air operation, approach chart, aerodrome map, en-route chart, Jeppesen, geo-database, geospatial-database, avionic, aeronautical navigation, GIS, CIS, flight planning, flight management.

Abstract

The key elements of geospatial data for use in military aviation include a variety of aeronautical charts. The following charts are those most commonly used: the Operational Navigational Chart, the Tactical Piloting Chart, the VFR charts following either ICAO requirements or commercial Jeppesen formats, the comparable national aeronautical charts with a 1 : 500 000 scale, the high resolution joint operations graphic maps or various specialized aerodrome maps and approach charts. Long distance operational planning and en-route navigation can be carried out using generalized IFR charts produced by commercial and governmental organizations.

In addition to aeronautical charts, there is a vast amount of geospatial data which are necessary for air operations and is provided in tabular form such as: location indicators, radio navigation aids, air space information and obstacles. Due to the high degree of technological integration in avionics, an increasing amount of this geodata is available in digital format on moving map systems or in integrated navigation and communication modules such as in a flight management system. This includes remote sensing, integration of global position systems, satellite imagery and simulation. To optimize air space use and mission efficiency while minimizing risk of accident, energy and emphasis must be focused on generating and providing up-to-date charts and critical geospatial data for specialized information systems.

1   Introduction

Five unique properties should be considered, when applying geospatial data and charts in the field of aviation:

> - Type of air space user,
> - Third dimension (z),
> - Distances,
> - Actuality and authenticity and
> - Level of IT and GPS integration.

Each one of these special traits influences air operations and affects the way geospatial data is handled (see International Civil Aviation at URL 7, Flight Services Jeppesen Sanderson Inc at URL 8 and Aviation System Standards National Aeronautical Charting Office at URL 9).

1.1   Air Space Use

With the exception of restricted or prohibited areas, all air space is constantly co-used by civil and military aircrafts. Civil aviators range from Para gliders to the transcontinental jumbo-jets. Military users include among others ground based air defense units, drones and reconnaissance teams, low flying army helicopters, slow and fast fixed wings aircrafts, tankers and AWACS following special pattern and strategic air lift. These aircrafts have substantially different air speeds, maneuverability and radar signatures. They regularly switch flight levels during the course of their missions. Due to the fact that air space is shared, interoperability for regulations, training, communications, technology and geospatial data is necessary for success. Two reactions can be observed:

> Aeronautical information and charts are produced and provided by civil authorities, specialized companies and military authorities which work in accordance and

> A substantial level of standardization is required by the ICAO in many areas of aerial operations.

This includes the uniform use of the WGS84 (World Geodetic System 1984) and the geographic degree/minute/second grid and reporting format, which is sometimes supplemented with the UTM (Universal Transversal Mercator) grid. As yet not resolved are the discrepancies between the metric and English systems (feet and miles versus meter and kilometer).

1.2 The z-Dimension in Air Space

Aircrafts operate in assigned air space boxes with specific height limitations or flight levels. In order to properly delineate the air space on the charts, not only are the geographic limits of the ATS (Air Traffic Services) borders required, but also accurate height information for each categorized box. Aerial operations can only be properly planned and executed if all necessary air space data is represented in three dimensions (Figure 3). The application of the third dimension on the geospatial data results in an increase in complexity (from the simple X-Y plane of most ground maps to a three-dimensional X-Y-Z space).

1.3 Distance

Due to the long distances covered by an aircraft during a single mission, different types of maps have been developed. These include detailed maps for takeoff and landing as well as for the mission areas, and more generalized maps for en-route navigation (Figure 4). In the case that more than one country is spanned during a mission, international diplomatic regulations for transit of military aircrafts must be respected and followed.

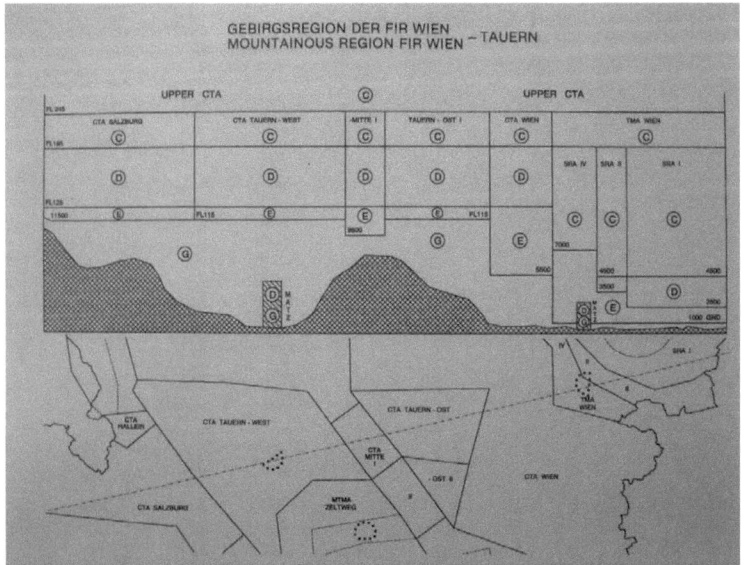

Figure 3: A simplified cross section of the FIR (Flight Information Region) WIEN, Austria. The top segment of the diagram shows a typical example of a three dimensional air space with designated boxes and the associated respective regulations; the lower part of the figure displays the conventional two-dimensional view of the map (data from the Austrian military Aeronautical chart with a scale of 1 : 500 000, Teichmann 2006).

1.4 Actuality and Authenticity

Air space information and air operation regulations are subject to short term change. Outdated aeronautical tables and charts provide pilots with faulty information and thereby increase safety risk factors. As a result, aeronautical data and charts need to be revised on a regular basis (i.e. tabular geodata are revised in Austria on a monthly basis, general charts on a yearly basis). Only authorized and verified aeronautical information should be used.

1.5 Information Technology

Since the beginning of flight, avionics has been on the cutting edge of technological innovation and of scientific development. Satellite communications, data links, GPS-navigation systems, digital moving maps and flight management systems are applied to increase flight safety and optimize air operations. In order to further streamline the efficiency of these modern technologies the available basic geospatial data has to be constantly integrated into various databanks. Modern air operations are currently in a transformation process, changing from operations with analogue charts and tables, to digitized databases with integrated command and control systems.

2 Aeronautical Charts used during Flight

Similar to nautical charts, aeronautical charts or aerial maps are primarily optimized for navigational application. The following charts are usually consulted during a flight mission (Figure 4). For takeoff and landing procedures special aerodrome and approach charts are needed. Flight navigation can be carried out using the ICAO charts or national VFR charts (Visual Flight Rules). For IFR (Instrument Flight Rules), generalized en-route charts are used. In addition, the ONC and TPC series provide common map images for long distance international flight navigation and planning. The JOG (Joint Operations Graphic) series could be the map of choice when operating an aircraft close to ground or during a joint mission. Detailed geoinformation about a target area is deducted from standardized high resolution ground maps in the range of 1 : 25 000 to 1 : 50 000. Depending on the purpose of a mission, even higher resolution geospatial data like aerial pictures or target folders are used. In addition, the standardized ICAO (International Civil Aviation Organization) chart is definitely the aeronautical and cartographical common denominator between a pilot in the air and his supportive elements like operational control or civil air traffic. Based on whether the flight is IFR or VFR and the category of air space used, an aviator and the attached units select the appropriate aeronautical chart (Figure 3 and 4, see also Aviation Publication Service Del Mar at URL 10, Maps, Globes, Travel

Guides Map Town Ltd at URL 11 and Maps & Travel books from World of Maps World of Maps at URL 12).

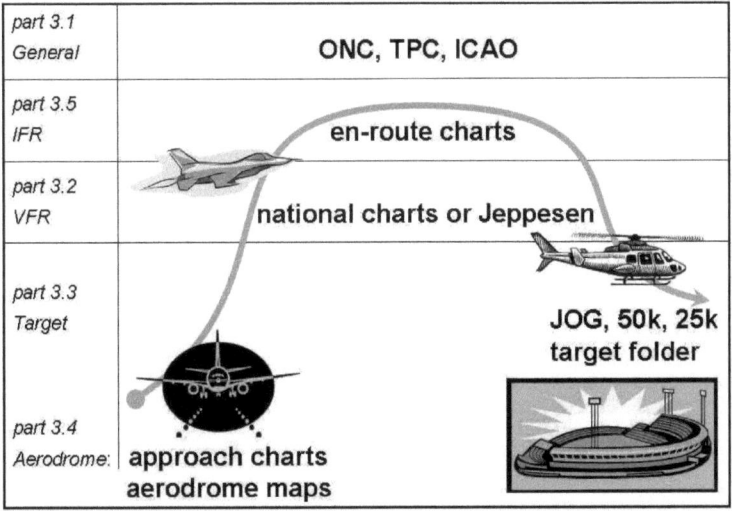

Figure 4:   A possible sequence of appropriate aeronautical charts for a given flight: First the approach and departure charts of the aerodrome, followed by IFR or VFR charts for the long leg and for the target area high resolution charts or images (Teichmann 2006).

3   Aeronautical Charts

In order to produce or choose an ideal chart for a mission, the above discussed special aeronautical requirements (i.e. different users of the air space, the third dimension, large distances, high actuality and the IT integration) must be considered. Natural and/or man-made features and aeronautical information are the main categories of information provided on any aeronautical chart. The key variations between the different aeronautical charts types are: the degree of generalization, the scale, the amount of information, its attributes and focus (see Maptech MapServer Aeronautical charts at URL 13 and National Geospatial-

Intelligence Agency Products and services at URL 14). For close ground operations, a high degree of information about natural (i.e. geomorphology) and man-made features (i.e. road-network) is necessary. However, for approach charts air space limitations and navigational aids are crucial. During en-route navigation the ease in reading the route vectors and navigation points on an IFR chart is more important than features on the ground. As a result, a large variety of aeronautical charts are available and each type of chart is optimized for a specific purpose. The following five subgroups of aeronautical charts are currently used for military planning and air operations (see Friebe Flugbedarf Luftfahrtkarten, manuals, pilot kits at URL 15, GCH Services, LCC – The Pilot Shoppe Aeronautical Charts at URL 16, Natural Resources Canada Aeronautical and Technical Services at URL 17 and MapTrax Australia Aeronautical Charts at URL 18). The aeronautical charts and data examples used in this article are primarily focused on the European scenario.

3.1   ONC, TPC and Similar Chart Series

The ONC (Operational Navigational Chart) and the TPC (Tactical Piloting Chart) series are government produced (primarily by the USA and UK) and cover most of the world's land mass. Both chart types have a Lambert Conformal Conical Projection, display a geographical as well as a UTM grid and are referenced to the WGS84. The multi-color print exhibits detailed geospatial and topographical information including: borders, urban areas, road and rail networks, vegetation, rivers and water ways, as well as detailed elevation information (Figure 5). In addition, both charts present extensive aeronautical information about aerodromes, vertical obstructions, radio facilities and air space categories or boxes. The ONC chart has a scale of 1 : 1 000 000 and is divided into 4 TPC charts with a more detailed 1 : 500 000 scale.

The following three charts series are concurrent to the ONC and TPC series:
  ➢ JNC (Jet Navigation Chart) with a 1 : 2 000 000 scale
  ➢ GNC (Global Navigation Chart) with a 1 : 5 000 000 scale
  ➢ WAS (World Aeronautical Chart) with a 1 : 1 000 000 scale.

These charts may serve as supplements or substitutes for the ONC and TPC series and are used by both military and commercial pilots.

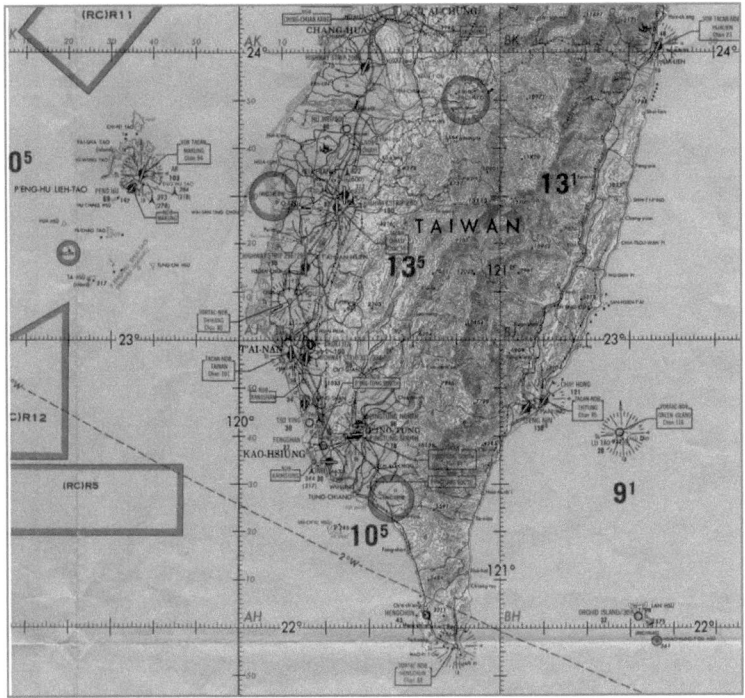

Figure 5: A section of the ONC chart covering southern Taiwan (URL 19), displaying relatively detailed geographical as well as comprehensive aeronautical information (www.lib.utexas.edu/maps/middle_east_and_asia/taiwan_onc_84.jpg; 22 april 2006).

The linked ONC and TPC series have advantages due to the fact that they provide almost complete coverage of most land areas, detailed topographic information, corresponding scales, double grids and comprehensive aeronautical information. Reliable update and distribution services for these chart series are carried out regularly. Therefore, the ONC and the TPC are excellent choices of basic charts for most international military operations, including planning and operational control (see Civil Aviation Authority Aeronautical Charts at URL 20, Federal Maps Inc Aeronautical Charts at URL 21 and Civil aviation administration Aeronautical Information Services at URL 22).

3.2   VFR Charts

Currently, the most common scale for VFR charts is 1 : 500 000 (see Federal Publications Inc Maps and Charts Aeronautical Charts at URL 23 and Map Connection Ltd Aeronautical at URL 24). A standardized ICAO chart printed with this scale is produced for most countries by their respective aviation authorities and usually has a regular updating schedule. A chart of this type is in most cases the authorized base for all air operations and is produced using a Lambert Conformal Conic Projection with a geographic degree grid and a scale of 1 : 500 000. The topographic information is provided in multi-color print with current aeronautical information overprinted in a unique dominant color like blue, purple or red. In addition to ICAO charts, similar Sectional Aeronautical Charts are available, which focus on the North American air space (see Pilotfriend Pilot resources directory at URL 25 and Mypilotstore Jeppesen and NACO Charts at URL 26).

Jeppesen also provides a highly valued commercial product with a 1 : 500 000 scale (Figure 6a). The VFR+GPS Jeppesen chart is available for many countries as well as regions and is also based on a Lambert Conformal Conic Projection with a degree grid, topographic and aeronautical information. The Austrian military flight chart 1 : 500 000, which is an example of a national VFR chart, is shown for cartographic comparison (Figure 6b). This chart also uses a Lambert Conformal Conic Projection with a degree grid and exhibits geographic, topographic and aeronautic information.

Figure 6:   Two examples for VFR charts. The two VFR charts are produced by different aviation authorities and show unique cartography. Despite a high degree of standardization and identical scale, each chart presents the necessary geodata differently.

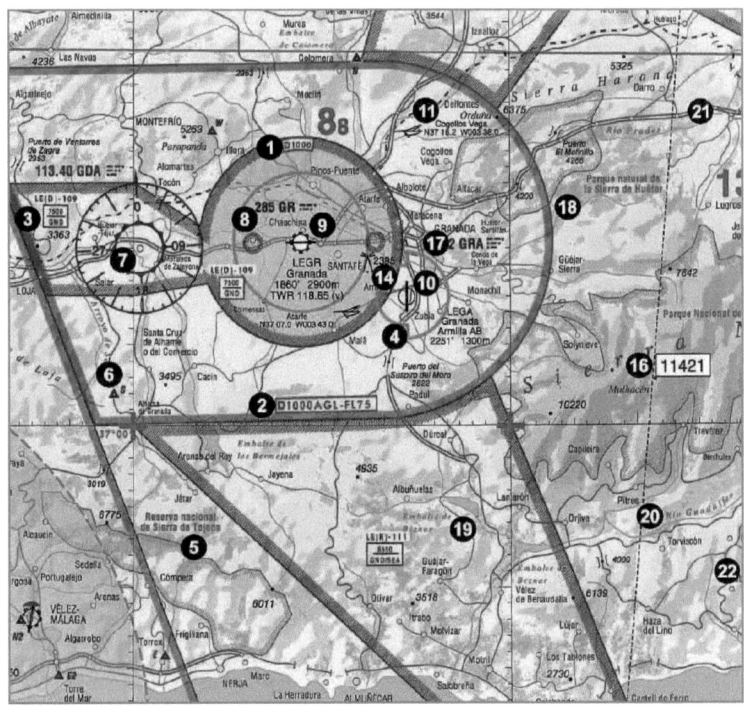

Figure 6a: Jeppesen VFR+GPS (URL 8) of the area Grenada, Spain (http://www.jeppesen.com/download/bottlang/VFR+GPS_combo.pdf;   22   april 2006).

Figure 6b: Austrian Military Aeronautical Chart 1 : 500 000 from the area
around Salzburg, Austria (Teichmann 2006).

3.3   JOG and Large Scale Maps

The JOG map series with a scale of 1 : 250 000 was primarily developed to
provide the different services (e.g. Army and Air Force) with a common unified
map picture and reporting grid (WGS84, UTM). Although there is some thought
that not enough geoinformation is provided for close ground operations on this
map type, it is the map of choice for joint operations.

Additionally, more detailed ground information can be obtained using regular
topographic maps like 1 : 25 000 or 1 : 50 000 or large scale road maps. For
special operations, more precise details for aviators could be accessed through
aerial photographs or enhanced target folders. Thus a common air and ground
operations map (JOG) in combination with other available geospatial data yield
the optimal result for aviator and ground troops (see National Geospatial-
Intelligence Agency Products and services at URL 14).

3.4   Approach Charts and Aerodrome Charts

Specialized maps and charts of each airport, aerodrome or heliport must be available for the flying community (see Airplane flight equipment Charts at URL 27, Softchart Aeronautical Products at URL 28 and Marv Golden Pilot Supplies Charts at URL 29). These are necessary for the approach and departure phases, during landing and takeoff as well as for maneuvering on the ground. These charts adhere to ICAO standards, including the necessary revision procedures, and are provided by national aeronautical authorities or by specialized civil companies like Jeppesen (Bottlang Airfield Manual).

For a typical minor international airport following types of geospatial data might be available in tabular form:

  ➢ Detailed location information about aprons, taxiways and check points
  ➢ Physical characteristics and distances of runways
  ➢ Helicopter landing areas
  ➢ Air space structure for the Control Zone
  ➢ Radio navigation and landing aids
  ➢ Special local procedures for IFR and VFR.

In addition to the above listed geospatial tables, the following charts and maps may also be provided with the following layout (color, scale), which could vary from airport to airport or depend on the chart producer:

  ➢ Aerodrome Charts (two-color production or print with different grey shades; 1 : 20 000, Figure 7c)
  ➢ Aircraft Parking Charts (two color, 1 : 4 000)
  ➢ Aerodrome Obstacle Charts (type A: two-color, 1 : 4 000; type B: multicolor, 1 : 25 000)
  ➢ Precision Approach Terrain Charts (b/w, 1 : 5 000)
  ➢ Standard Instrument Departure Charts (b/w, 1 : 500 000)
  ➢ Standard Instrument Arrival Charts which are cartographically similar to the Standard Instrument Departure Charts
  ➢ Instrument Approach Charts (two-color, 1 : 250 000; Figure 7b)

- Visual Approach Charts (multicolor, 1 : 250 000; Figure 7a)
- TMA Charts (two-color, 1 : 500 000)
- VFR Charts (multicolor, 1 : 250 000)
- Terminal Aeronautical charts (multicolor; 1 : 250 000).

This lengthy chart list illustrates the complexity and breadth of the geospatial data necessary for all aspects of aviation close to the airfield (see ICAO International Civil Aviation at URL 7 and Flight Services Jeppesen Sanderson Inc at URL 8).

Figure 7:   Examples of Aerodrome Navigation Charts (URL 30) for the airport Antwerpen, Belgium (for simulation use only – www.ivao.org.be).

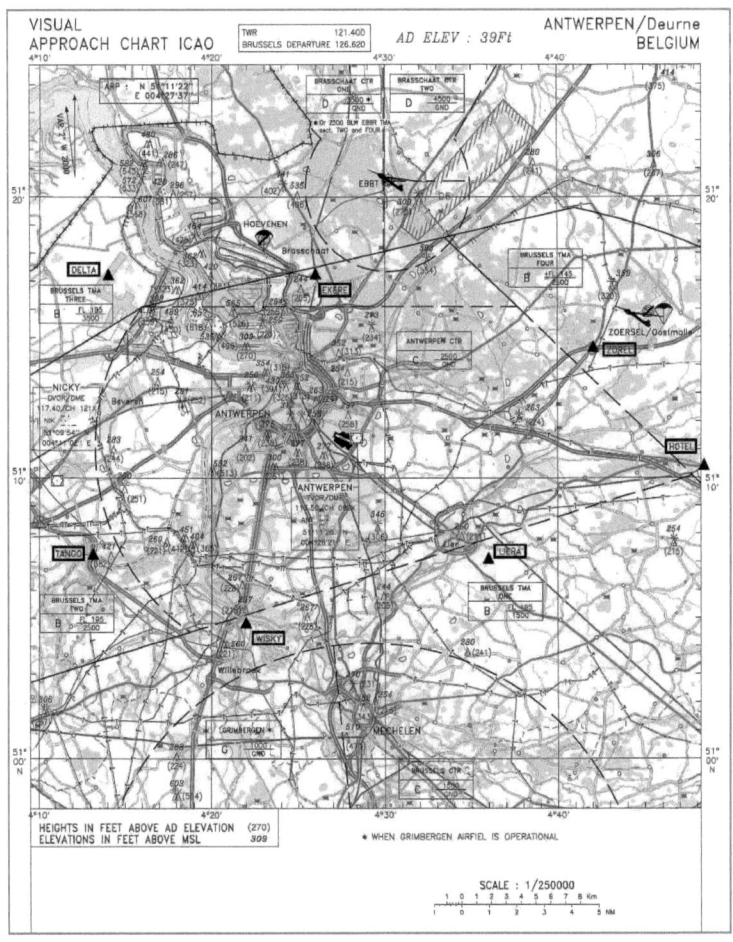

Figure 7a: Visual Approach Chart
(http://www.ivao.be/files/charts/Civil/EBAW/EBAW%20VAC%2001.pdf; 22
april 2006).

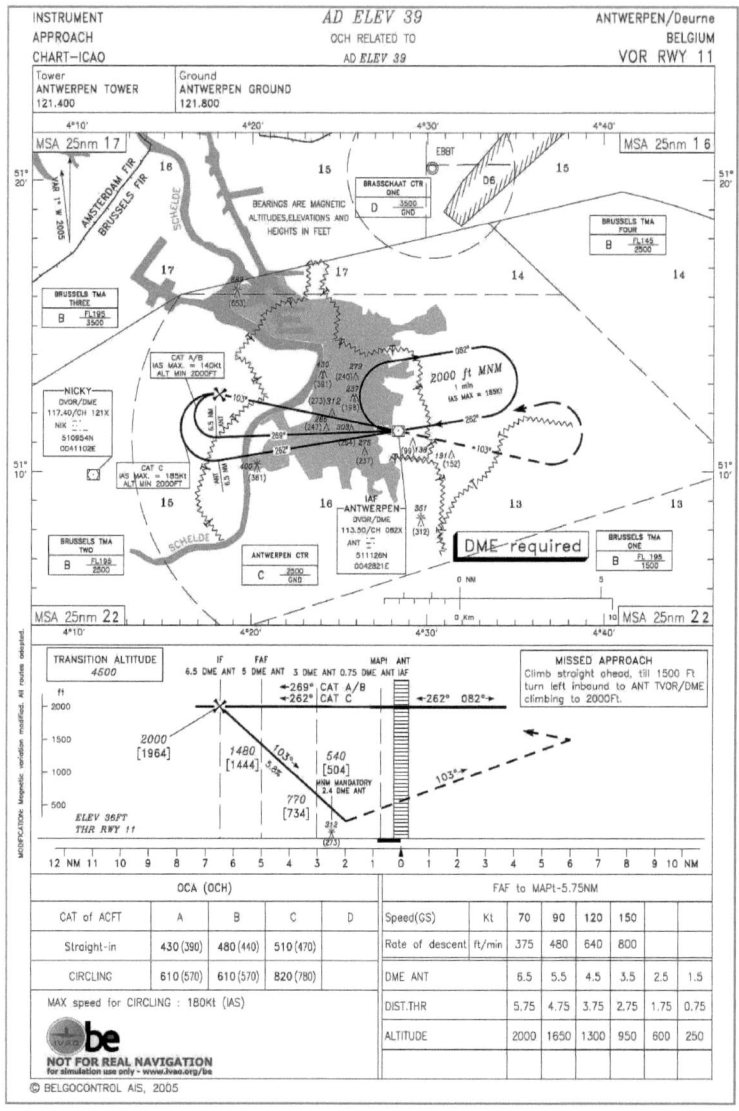

Figure 7b: Instrumental Approach Chart

(http://www.ivao.be/files/charts/Civil/EBAW/EBAW%20IAC%2003.pdf; 22 april
2006).

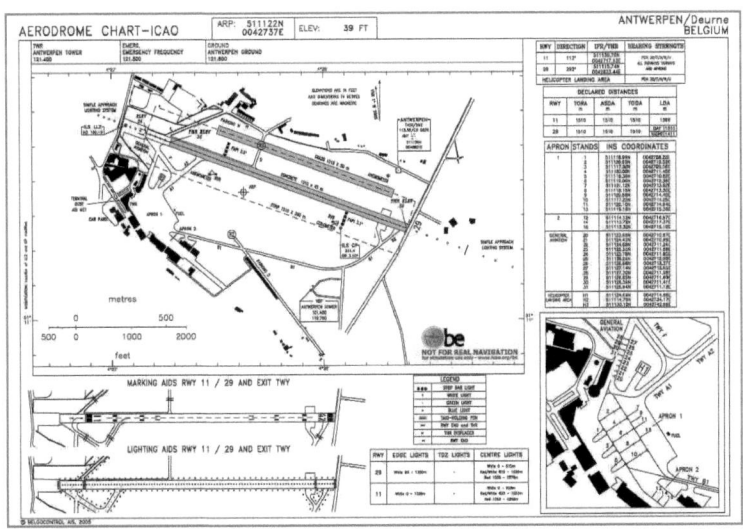

Figure 7c: Aerodrome Chart
(http://www.ivao.be/files/charts/Civil/EBAW/EBAW%20ADC%2001.pdf; 22
april 2006).

3.5   En-Route and Small Scale Charts

Specialized small scale aeronautical maps or IFR charts for long distance
operations and planning are available, among others, through the following
institutions: Eurocontrol, US Department of Defense (DOD), National
Aeronautical Charting Office and Jeppesen (see Jeppesen – Making every mission
possible at URL 4, National Geospatial-Intelligence Agency Products and services
at URL 14 and Eurocontrol Cartography at URL 31). Eurocontrol produces the
Airspace Management Planning Charts (ASM) with a scale of 3,5 Mio, Lambert
Conformal Conic Projection and a geographic degree-grid.   In addition,
Eurocontrol issues a similar chart, the Central Flow Management Unit Chart
(CFMU) with a scale of 3 Mio (Figure 8). The DOD issues the Flight Information
Publication En-route High Altitude and the En-route Low Altitude Charts for
extensive areas. Jeppesen also produces High Altitude En-route Charts, as well as
Low Altitude En-route Charts.   These IFR en-route charts are network maps

which focus on nodes and vectors and are used for long distance navigation and operational planning. In addition, several small scale charts and maps for specialized aerial use are available: the Chart of the Aerodromes, the Pilots World Satellite Wall Map and the Free Flight Atlas as well as proprietary en-route charts of several commercial airlines.

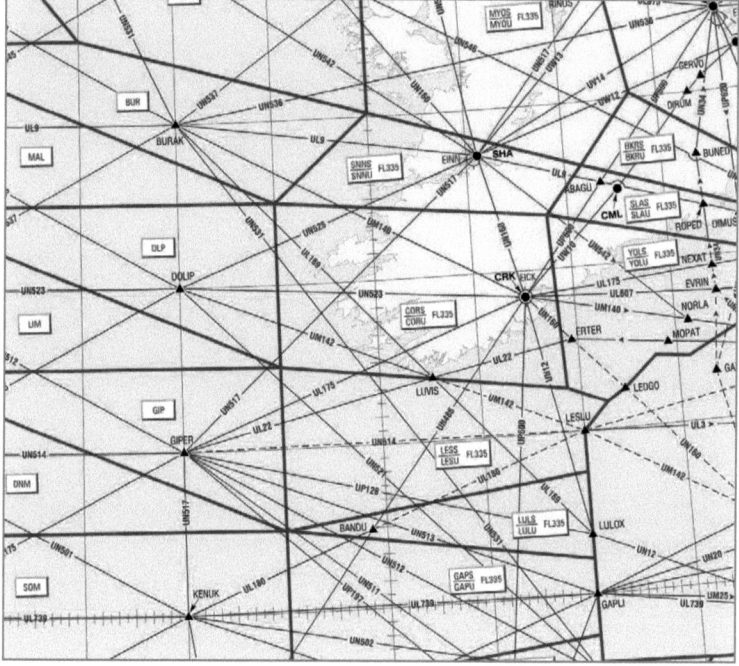

Figure 8: IFR-chart example: Central Flow Management Unit Charts (CFMU Charts) map of Eurocontrol (URL 31) showing the southern part of Ireland; focus is on air space limitations, the vectors (routes, corridors, etc.) and the nodes (waypoints, navigation aides, etc.)
(http://www.eurocontrol.int/carto/gallery/content/public/documents/Updates/UL1 69_16mar2006_HiRes.tif; 22 april 2006).

4   Additional Geospatial Data for Aviation

The aeronautical chart is one of the most prominent and indispensable representations of geographical information for aviators, but by far not the only one. In addition to the above described aeronautical charts a whole range of other geodata is used extensively in aviation (see Jeppesen – Making every mission possible at URL 4, ICAO International Civil Aviation at URL 7, Aviation System Standards National Aeronautical Charting Office at URL 9 and Eurocontrol Cartography at URL 31).

4.1   Location Indicator

The most dominant of these is the global standardized location indicator. The four letter code is ICAO standardized and is derived hierarchically: first letter region, second letter country, third letter type of location/airfield, fourth letter specific designator. For example, LOXT is translated as: L for southern Europe, O for Austria, X for military airfield and T for Tulln-Langenlebarn.

4.2   Air Space Structure

Graphical representation of air space (Figure 3) is derived for each individual country from an authorized tabular publication. The regulated and geographically defined air space of the Air Traffic Services (ATS) includes primarily the Flight Information Region, Terminal Control Area, Aerodrome Traffic Zone, Control Zone, prohibited, restricted and danger areas as well as regions for sport and recreational activities like gliding areas. In addition to these different ATS air space categories, the locations of lower and upper navigation routes are also published in geographic coordinates.

2.1	FIR, UIR, ACC AOR, UTA, CTA, Deligation of the Resposibility for Provision of ATS and TMA		
Upper and lower Flight Information Regions		**ATS Airspace Classification. Ref ENR 1.4**	
Name and lateral limits		Vertical limits	ATC unit and callsign
SWEDEN FIR 690336N 0203255E–Swedish/Finnish border Southward to 653148N 0240824E–644100N 0225500E–633700N 0213000E– 632830N 0204000E–631000N 0201000E–614000N 0193000E– 610000N 0191905E–801803N 0190756E–801130N 0190512E– 593346N 0195859E–591524N 0203239E–590000N 0210000E– 573410N 0200900E–570000N 0195000E–555100N 0173300E– 545500N 0155200E–545500N 0150807E–clockwise along an arc of 16.2 NM (30 km) radius centred on 550404N 0144448E– 545500N 0142127E–545500N 0125100E–552012N 0123827E– Swedish/Danish border northward to 561253N 0122205E– 583000N 0103000E–584540N 0103532E–585332N 0103820E– Swedish/Norwegian border to 690336N 0203255E		FL 285 / 8700 m STD GND	See ACC AOR below

Table 1:   Example of the geographic definition of the Flight Information Region (FIR) Sweden (URL 32); displaying:

> Geographic coordinates of the polygon describing the extent of the area
> Vertical limits of this specific air space box (e.g. from GND to FL 285)
> Classification of this air space and additional information pertaining to this air space like the radio call sign

(http://www.lfv.se/upload/ANS/AIP/ENR/ENR%202/ES_ENR_2_1_en.pdf;   22 april 2006).

4.3   Radio Navigation Aids

A second very important database of geographical reference points in an aeronautical publication is the table of radio navigation aids. This standardized table describes the type and name of the station, an identification code, the frequency, the hours of operation, the x-y-z-coordinates of the radio navigation point and additional remarks. It is supplemented by various databases with name-code designators for significant points which can be selected to be waypoints of a flight and navigational system.

Station	Service	ID	FREQ	HOR	Position GEO	Situation / ALT ft Location / Alt	exploitant operator	Observations Remarks
ABBEVILLE	VOR-DME	ABB	108.45 MHz CH21Y	H24	50°08'06.5"N 001°51'16.9"E		AVA	Portée 60-500
AGEN-Gaudonville	VOR-DME	AGN	114.8 MHz CH95X	H24	43°53'16.9"N 000°52'22.3"E	ALT 896ft	AVA	Portée 120 VF/100-50
AJACCIO-Coti Chiavari	VOR-DME	AJO	114.8 MHz CH95X	H24	41°46'13.9"N 008°46'28.8"E 41°46'12.3"N 008°46'28.1"E	ALT 2142ft	AVA	Portée 200 GS/100-50
AMBERIEU	TACAN	AMU	CH110X	H24	45°59'19.0"N 005°19'52.5"E	ALT 820ft	FAF	Portée 120NR/100-50.
AMBOISE	NDB	AMB	341 kHz	H24	47°25'05.4"N 001°02'27.6"E		AVA	PORTEE 50
	VOR-DME	AMB	113.7 MHz CH84X	H24	47°25'44.1"N 001°03'52.0"E	ALT 387ft	AVA	Portée 80-50
ANGERS	VOR	ANG	113 MHz	H24	47°32'12.7"N 000°51'06.6"W		AVA	Portée 120 VD/80-50
AUTUN	VOR-DME	ATN	114.9 MHz CH96X	H24	46°46'21.4"N 004°15'32.9"E 46°48'19.7"N 004°15'33.8"E	ALT 686ft	AVA	Portée 120 TX/60-50
AVIGNON CAUMONT	L	CM	389 kHz	H24	43°54'26.8"N 004°54'19.4"E	133°/487m DTHR 17	AVA	
AVIGNON Pujaut	VOR	AVN	112.3 MHz	H24	43°59'43.3"N 004°44'47.0"E		AVA	
AVORD	NDB	AVD	288.5 kHz	H24	47°07'14.4"N 002°47'58.6"E		FAF	
	TACAN	AVD	CH43X	H24	47°03'29.1"N 002°37'47.5"E	ALT 597ft	FAF	Portée 150-60
BALE-MULHOUSE	VOR-DME	BLM	117.45 MHz CH121Y	H24	47°37'58.1"N 007°29'58.2"E	336°/1NM THR 16 ALT 896ft	AVA	Portée 40-25
BASTIA PORETTA	VOR-DME	BTA	114.15 MHz CH88Y	H24	42°34'24.9"N 009°28'29.4"E	343°/1175m THR 16 ALT 36ft	AVA	Portée 80-50
BIARRITZ BAYONNE ANGLET	VOR-DME	BTZ	114.15 MHz CH88Y	H24	43°27'59.9"N 001°30'37.0"W	201°/223m THR 27 ALT 276ft	AVA	Portée 100 VD/80-50

Table 2: Example of En-Route Radio Navigation aids with station ID, service, operational information and geographic position, including altitude; example for France (URL 33)

(http://www.sia.aviation-civile.gouv.fr/aip/enligne/METROPOLE/AIP/ENR/4/AIP%20FRANCE%20ENR%204.1.pdf; 22 april 2006).

4.4 Air Navigation Obstacles

Another key database of aeronautical geoinformation is the table of air navigation obstacles. When a mission has to be carried out close to ground, it is essential for a pilot to be fully aware of the location of any obstacles. This air navigation obstacle database should be available both on board an aircraft (i.e. in a digital map system or flight management system) and on ground during mission planning and execution in order to maximize awareness and minimize risk.

Standort Location	Art Type	WGS 84 Coordinates WGS 84	GND Height above GND	Fußpunkthöhe ELEV	zeichnung Day marking	Befeuert Lighted
1	2	3	4	5	6	7
NIEDERÖSTERREICH:						
Bezirk Hollabrunn:						
Windpark Guntersdorf Ost - Anlage 1	Windkraftanlage Windpower plant	N48 40 18 E016 04 02	121M / 397FT	274M / 899FT	ja / yes	ja / yes
Windpark Guntersdorf Ost - Anlage 2	Windkraftanlage Windpower plant	N48 40 27 E016 04 05	121M / 397FT	283M / 928FT	ja / yes	ja / yes
Windpark Guntersdorf Ost - Anlage 3	Windkraftanlage Windpower plant	N48 40 17 E016 04 24	121M / 397FT	279M / 915FT	ja / yes	ja / yes
Windpark Guntersdorf Ost - Anlage 4	Windkraftanlage Windpower plant	N48 40 27 E016 04 27	121M / 397FT	282M / 925FT	ja / yes	ja / yes
Windpark Guntersdorf West - Anlage 1	Windkraftanlage Windpower plant	N48 40 31 E016 01 42	121M / 397FT	273M / 898FT	ja / yes	ja / yes
Windpark Guntersdorf West - Anlage 2	Windkraftanlage Windpower plant	N48 40 18 E016 02 30	121M / 397FT	276M / 906FT	ja / yes	ja / yes
Windpark Guntersdorf West - Anlage 3	Windkraftanlage Windpower plant	N48 40 29 E016 02 33	121M / 397FT	282M / 925FT	ja / yes	ja / yes
Windpark Guntersdorf West - Anlage 4	Windkraftanlage Windpower plant	N48 40 18 E016 02 54	121M / 397FT	276M / 906FT	ja / yes	ja / yes
Windpark Guntersdorf West - Anlage 5	Windkraftanlage Windpower plant	N48 40 29 E016 03 00	121M / 397FT	281M / 922FT	ja / yes	ja / yes
Bezirk Melk:						
Pöchlarn	Windkraftanlage Windpower plant	N48 10 47 E015 13 18	124M / 407FT	338M / 1109FT	ja / yes	ja / yes
Bezirk Wr. Neustadt:						
17/100/5087 NOWB 049 Rohr im Gebirge/Rohrer Sattel tele.ring Telekom Service GmbH	Sendemast / Radio mast	N47 52 12 E015 47 48	32M / 105FT	870M / 2854FT	nein / no	nein / no
Bezirk Waidhofen an der Ybbs:						
19/100/8300 NOWY 008 Waidhofen an der Ybbs/ Hochkogel tele.ring Telekom Service GmbH	Sendemast / Radio mast	N47 56 56 E014 41 30	36M / 118FT	810M / 2657FT	nein / no	nein / no

Table 3: This section of an air navigational obstacle table from Austrocontrol for the state of Burgenland in eastern Austria displays the following parameters for each location: type, coordinates, height above ground, elevation and type of marking (URL 34)
(http://www.austrocontrol.at/multimedia/LO_SUP_2005_02_en_tcm586-48394.pdf?frame=leftnav; 22 april 2006).

5 Information Technology Integration

Under pressure to minimize equipment size, standardize and increase effectiveness while maintaining flight safety, aviation industries have quickly embraced new Communication and Information Systems (CIS) technologies. The need to adopt CIS technology is also driven by the constant demand to accelerate the decision making process (i.e. shorten the reaction time) and to optimize air operations (i.e. fuel efficiency). Air operations planning, dispatch or ground control teams can utilize Geographical Information Systems (GIS) work stations to update, provide and network all necessary geospatial data for use in aeronautical operations (Figure 9). Such a system might also be applied for complex evaluation during simulation runs in order to optimize mission efficiency. On board modern airplanes, specialized CIS systems could constantly provide important geoinformation during flight. The CIS system on board generally includes a digital map system (Figure 9b) which can store a wide range of charts (e.g. VFR, ICAO or JOG) and has a large range of functions (i.e. zoom, distances, switch charts, waypoints). Aeronautical databases can be incorporated into modern flight management systems or into highly automated communication and navigation modules and mission planning systems (figure 9c). The push to increase the integration of modern technologies to update current avionic systems also includes the task of providing optimal geographic data, using high resolution imagery or satellite data.

Figure 9:   Digital airspace charts and aeronautical GIS.

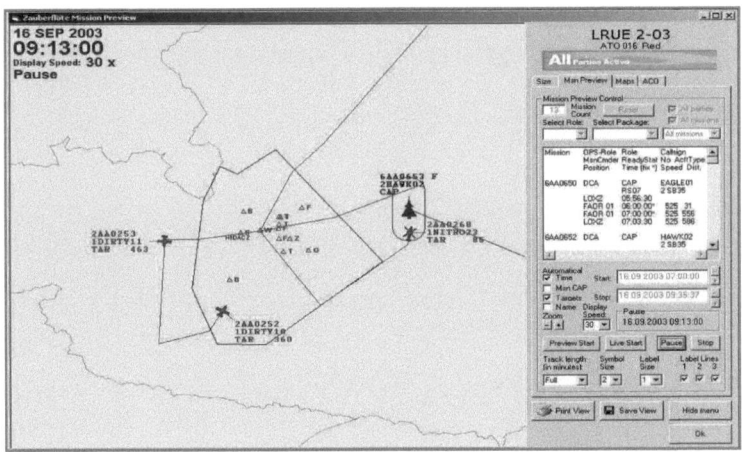

Figure 9a: A screenshot from the Austrian Air Force Command and Control System *Zauberflöte* (magic flute), currently under development (status as of 2006, Teichmann 2006).

Figure 9b: A picture of the digital moving map system installed in an Austrian Air Force S-70 Black Hawk Helicopter (Teichmann 2006).

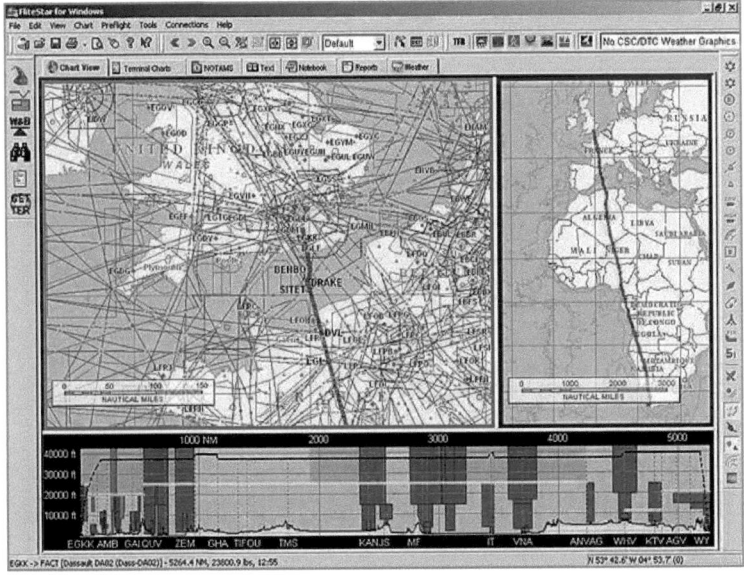

Figure 9c: A screenshot of *Jeppesen FliteStar for Windows* program (URL 4)
(http://www.jeppesen.com/wlcs/commerce/catalog/staticFile.jsp?fileName=includ
es/FliteStar9-4.html; 22 april 2006).

6   Conclusions

Unique situations require special measurements. Different airspace users, three-dimensionality of air space, large distances, actuality and IT-integrations provide substantial challenges to aviators and ground support elements. In order to overcome these challenges, an ever increasing amount of detailed and standardized geographical information must be generated and made available to the aviation community. These geographical data include highly specialized aeronautical charts which range from large to small scale and IFR to VFR. The charts are supplemented by tables of geographical reference points. The proper maintenance and usage of these two geographical information types in conjunction with modern technology increases flight safety, mission efficiency and reduces risk factors.

7 Glossary

AIP	Aeronautical Information Publications
AIS	Aeronautical Information Services
ATS	Air Traffic Services
AWACS	Airborne Warning and Control System
CIS	Communications- and Information System
GIS	Geographic Information System
ICAO	International Civil Aviation Organization
IFR	Instrument Flight Rules
JOG	Joint Operations Graphic
ONC	Operational Navigational Chart
TPC	Tactical Piloting Chart
VFR	Visual Flight Rules

8 Bibliography

See comprehensive summary at the end

9 Author

LTC MSc MAS Mag. Dr. Friedrich TEICHMANN

1964:	Born in Salzburg, Austria
1988:	M.Sc. in Geological Sciences, University of Maine, USA
1989:	Mag. rer. nat. in Geology; University of Salzburg, Austria
1995:	Ph.D. in Environmental Sciences, University of Rochester, USA
1996:	Head of Military Geographic Services of the Second Army Corps, Austria
2000:	Head of Military Geographic Services of the Austrian Air Force
2001:	Postgraduate MAS in Geographic Information Sciences, University of Salzburg, Austria

2002:   Lecturer for GIS, University of Applied Sciences, Wiener Neustadt, Austria

2003:   Department Head of Communication and Information Systems, Austrian Air Force

PRESS RELEASE

Reinhard MANG and
Hermann HÄUSLER (eds.)

International Handbook
Military Geography

591 pages, 160 x 235 mm, more than 350 figures and 40 tables, hardcover

€ 35,-

€ 20,- by purchase of 10 handbooks or more
€ 15,- for authors

ISBN 3-901183-50-7
AV + Astoria Druckzentrum GmbH, Vienna 2006

At the beginning of the third millennium, classical Military Geography has changed dramatically. New challenges, worldwide scenarios for military forces, and an ever increasing international cooperation, including civil and non governmental organizations, require comprehensive applied geoscientific efforts. This Handbook is a first attempt to show the wide range of numerous military geographic activities concerned.

The handbook covers a great number of important topics - from the earth's shape to climate, from a basic theory to geopolitical aspects, from lessons learned in history to present and future geospatial requirements, from geographic reality to its models, such as maps, fact sheets, and so on.

Over 40 authors from seven Nations - each one a capacity in his profession - present Military Geography by a handpicked number of over 50 papers. Emphasis has been laid on an easy access for the reader by strictly limiting the papers in size and stressing easy and clear phrasing. Paper-related abstracts, glossaries, bibliographies, and an overall index allow for fast and deepened information gathering and thus guarantee a wide use of this book - for the military, the student, the geoscientist, for any kind of geospatial based decision maker and last but not least - for the public.

As long as military activities take place in the geographic space, Military Geography turns out to be an indispensable prerequisite for any military operation worldwide! This book will tell you why and will show you how.

Figure 10:   Press Release of the International Handbook Military Geography 2006, ISBN 3-901183-50-7

8 August 2006

Dear General Jilke,

Thank you for sending me a personalized copy of the *"International Handbook Military Geography"*. This book is an excellent resource and a wonderful addition to my library. I look forward to reading it at my earliest opportunity.

Please pass my thanks to the Brigadier Dr. Mang and University Professor, Dr. Hausler for their exceptional work. I was greatly honored to write the foreword for this book.

Thank you for your thoughtfulness and I wish you all best for continued success.

Most sincerely,

James L. Jones
General, U/S. Marine Corps

Major General Wolfgang Jilke
Austrian Military Representative to the
Military Committee of the EU, EAPC
NATO Headquarters
Boulevard Leopold III
B-1110 Brussels
Belgium

Figure 11:   Acceptance note of General James L. Jones, Supreme Allied Commander Europe, for the Handbook of Military Geography.

Part III

Challenges Facing System Integration and Network Focused Integration of the Geospatial Data

The current design challenges and the future research scope can be explained using a two stage model. The first stage is the optimal system integration or embedding of the geo-system in the avionic surroundings. The second step is the implementation of network based operation processes using the geospatial data. Several aspects of the implementation have to be recognized. Based on the fact, that most modern IT-implementations are complex and require additional resources, only implementations with well documented benefits and operational advantages should be carried out. These benefits might include: Reduced operational costs, more accurate data and therefore reduced error potential, critical time advantages, better situation awareness or improved flight safety. The following improvements in aviation operations would result in a distinctive competition advantage both in the military and in the civil domain. The necessary high levels of security and secrecy in this aspect have to be recognized and accepted. This is an additional reason, why the subsequent discussion focuses on the avionic system level and application architecture and does not go into detail about technical implementation.

1 System Integration

The initial point for integrating a geo-information-system into the avionic surrounding could be the functions required at the cockpit interface (Figure 12). The functions on the user interface should be optimized for simple selection in case of high urgency or emergency, when quick reaction is required. These rudimentary, minimum functions should include Zoom, Scale, Pan the map, Orientation of the display (head up or north up) and chart Selection. However, for complex missions or situations additional geospatial data is a necessity (Scott 2000, Steve 2003). Therefore, the geo-information-system with its user interface must be flexible enough to include complex GIS-analysis on its display (Schwarz 2004, Thomalla 2006).

The next considerations for integrating or embedding the geo-information-system in the avionic complex are the hardware constraints (Figure 12). The interrelationship between the avionic hardware used has to be optimized to be concurrent with the geospatial data that is used on board. One typical issue is the subject of pixel resolution used during the generation of the implemented raster graphic maps. An optimization process between the resolution of the user interface display and the raster chart resolution which includes the graphic format has to be carried out. This task immediately leads to questions about memory and computation speed of the onboard geo-information-system. Several aspects come to mind: The available memory space relative to the amount of geospatial data loaded, the programmed continuity of map sheets versus the computing power of the geo-information-system, as well as the overall system architecture of the digital avionic geo-system. These have to be optimized to fit the requirements of the geospatial data used.

The final aspect of an innovative digital geo-information-system on board is advanced GIS-functionality (Figure 12). Modern GIS-technology provides several advantages over original GPS-coupled raster map display systems. Various geospatial data can be integrated and loaded. This includes vector and raster maps as hybrid maps, digital elevation and digital terrain models, three dimensional analyses, extended access to geospatial databanks like tables, oblique aerial pictures, remote sensing data and results of location based services. These new functions permit on-the-fly advanced spatial analysis.

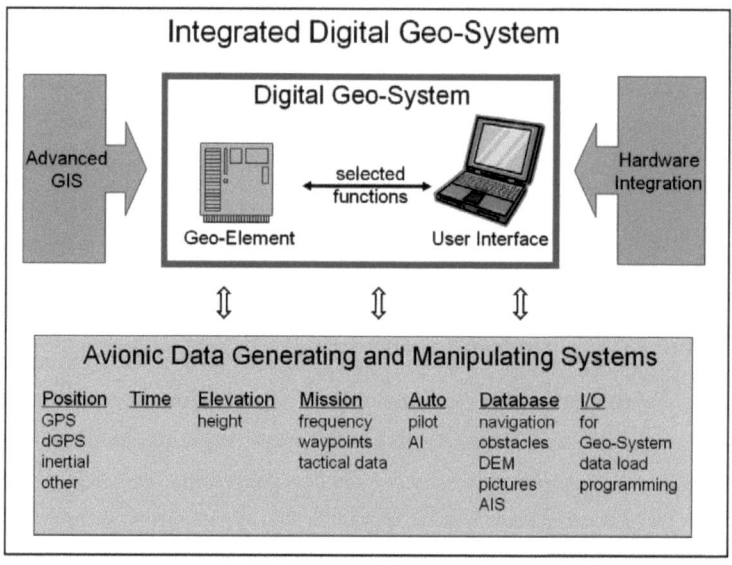

Figure 12:   Architecture of an Integrated Digital Geo-System (Teichmann 2007).

An advanced integrated or in the avionic environment embedded geo-information-system could have these connections and benefits (Figure 12). Standard positioning input could be upgraded to a higher resolution differential GPS input or could be supplemented with other positioning inputs. The elevation (height above ground) data should be coupled with a GIS elevation terrain model and could provide position verification and, if implemented, an advanced forward looking ground proximity warning. The combination of time and flight management data with the GPS track and the GIS-technology of the embedded geo-system would provide state of the art analysis functions. A geo-information-system linked to the flight management system could permit optimal data management and programming for all avionic systems and could also serve as the input/output unit for the geo-system. The geospatial data might include, in addition to those discussed in part II, an additional broad range of specialized data sets like DTED (Digital Terrain Elevation Data), DVOF (Digital Vertical Obstruction File), DAFIF (Digital Aeronautical Flight Information File), HIRTA (High Intensity Radio Transmission Area) as well as meteorological data (North 2002, Penney 2000, Warwick 2001). Based on the geo-referenced obstacle data or the DVOF on board, location based obstacle warnings could be generated by the embedded geo-system and thereby increase the safety of operations carried out close to the ground. Finally, an embedded or fully integrated geo-information-system would have access to all mission relevant data like frequency tables, ATS or AIS, contingency plans, holding patterns or weather information. This aspect alone might provide the greatest argument for the integration of the geo-information-system in the avionic environment.

New input and output options are currently being development for the cockpit. Examples are: direct-voice-input and head-up-display. These would combine nicely with the new integrated geo-information-system and its modern functionalities. Through this integration the situation awareness and the mission liaison could be immensely improved.

2   Network Focused Integration of Geospatial Data

The development of an integrated or embedded geo-information-system would be a great leap forward in avionics. An ultimate advantage of which would be experienced through the network focused application of relevant geospatial data. Based on the hypothesis that information superiority leads directly to competitive advantage, optimal location and communication information in such a highly mobile business as aviation must result in increased efficiency and effective output. Linking position, with or without additional geospatial data with voice or data communication, is one of the rapidly emerging markets. The following three paths should be analyzed for network focused operations (Penney 2001, Quaranta 2002). Figure 13 shows the secure LAN at the Air Base, the secure in-flight communication and data link to ground control and the open in-flight communication.

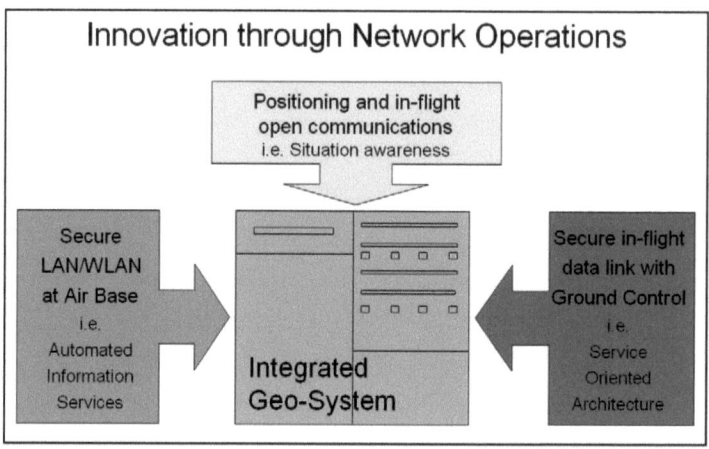

Figure 13:   Innovation for the integrated Geo-system through network operations (Teichmann 2007).

2.1   Secure LAN at the Air Operation Base

The most apparent improvement for a network focused application of geospatial data can be achieved through the installation of a secure data communication after the flight. Dependent on the individual situation and security regulations at a particular Air Base, that communication line could be wireless and highly mobile. The immediate improvement of this setup would be to install automated download and update procedures between the aircraft and its home Air Base. Option for the data download should include, among all the other technical reports, the GPS-track or any other mission information recorded during the flight. The upload from the ground station could include, from the geospatial data aspect, a whole range of information. Different types of geospatial data require different update cycles. These could range from daily updates for the obstacle files, to yearly updates for the raster maps. The above discussed secure LAN at the airbase could be applied not only for updating the relevant geodata, but also for transferring other relevant information, including mission and maintenance data.

An information service could be implemented, that would carry out all necessary updating procedures automatically. This would depend on individual requirements of the data set, update cycles and the links to a central data warehouse. The advantages of an automated information service by a secure data link on ground are clear. The on-the-fly automated network procedures for individual geospatial data sets, which are update by the end user with the central data warehouse, will result in faster turn around times and more reliable geospatial data available on board. This advanced network can also be applied to additional technical or mission data transfer.

2.2 Secure data–links with Ground Control

A second aspect of the network focused application of geospatial data is found in the communication between the aircraft in the air and the operational control center (Figure 13). This communication link has to cover the area of interest for operations and/or the aeronautical range and must enable both voice and data communication. If essential information is to be exchanged, a secure channel should be selected. This information exchange could provide the following new services.

An aircraft sends a standardized automated situation report with it's current coordinates. This permanently documented data would enable the ground operations center to apply all current and future telematic data applications. This would include services like fleet management and simulation scenarios among others. Additional benefits could result from direct coordination and cooperation between individual aircrafts of a fleet.

If a broadband communication channel is available as a secure data link, further applications might be possible. On demand location based services could be offered for the aircraft crew based on the constant accessibility of navigation, communication and geospatial information. Specialized service oriented application architecture could provide optimized information management. This might span from secure information regarding tactical mission data or flight plan modifications to inflight-to-ground video teleconferencing. Similar to the contemporary NATO data link 16, mission related geospatial data, like latest remote sensing data or radar data, could provide the additional critical time information needed to ensure a successful mission.

2.3   Open in-flight Positioning and Communication

The final network path is the open communication between the aircraft in the air to other elements (Figure 13). The purpose of this modern communication line is not only to improve the situation awareness in the air, but also to enable aeronautical passenger services. The first corresponding communication partner is the air traffic control authority. This contact is carried out by radio and focuses on navigation points and air traffic information services.

Open in-flight communication channels might also be used to provide differential GPS corrections and thereby increase the position accuracy. Another key application for open in-flight communications could be developed, if either the required transponders are modified or additional avionic communication equipment is installed. This equipment would then transmit the coordinates of the aircraft to anyone with a corresponding receiver. Such a system would link position with open communication and therefore increase situation awareness and reduce risk of air traffic accidents. This could be a supplement to or substitute for the TCAS (Traffic Alert and Collision Avoidance System), which is currently being installed in the larger aircrafts (URL 35). In addition to critical traffic information services, other major improvements are possible. These could be gained through use of standardized weather information provided in a network structure which would link ground stations and all participating aircrafts in one framework.

Conclusions

Processing any geo-data, including those used for aeronautics, from a geographer's standpoint would ensure optimal results, because a geographer has the training, means, ability and knowledge to bring together and consider all relevant aspects of a given geo-situation at once. The geographer is not interested in one aspect over another, but integrates all of the geospatial database information to generate an optimal geo-model, which can be used by the entire aeronautical community.

Geospatial data for aviation range from navigation charts to mission related way points on tables. The geo-information-system providing this data has evolved from mechanical devices into stand alone digital systems and further into integrated units.

The advanced geo-information-systems should be embedded in the avionic surrounding to allow for maximum data exchange. Modern digital geo-systems can provide a wide range of advanced GIS-functions. Linked together with a flight management system, extensive mission-critical spatial analyses could be carried out and recorded.

The procedures for utilization of the geospatial data should be networked. Secure LAN communication at an air base provides the base for an automated information service. Secure in-flight communications between an operation center and a pilot could evolve into an on demand service oriented application architecture and telematic applications. Open in-flight communications with a standardized protocol, which includes aircraft position, would increase the situation awareness and reduce risk for all flights. These new developments in communication and information services could be supplemented and enhanced through the implementation of the next IT-generation like semantic web. These could in turn be fully integrated with flight plan management and aircraft operations systems.

In short, network focused handling of all geospatial data for aviation would result in faster reaction times and improved safety. Intelligent application of geospatial data would ultimately reduce aviation operating costs.

References

<u>Monographs:</u>

Blanchard, W., 1995: The air pilots guide to satellite positioning systems. Airlife Publishing, Shrewsbury, UK.

Cayne B.S., 1993: New Webster's Dictionary and Thesaurus. Lexicon Publications, Danbury, USA.

Clarke B., 1998: Aviator's guide to GPS. McGraw-Hill, New York, USA.

Grewald M. et al., 2001: Global Positioning systems, Inertial Navigation and Integration. John Wiley & Sons, New York, USA.

Hake G., 1982: Kartographie. Sammlung Göschen de Gruyter, Berlin, BRD.

Hitchcock J.E., 1997: Navigation for pilots. Airlife Publishing, Shrewsbury, UK.

Hofmann-Wellenhof B. et al., 2003: Navigation – principles of positioning and guidance. Springer-Verlag, Vienna, Austria.

Leibbrand, W., 1984: Kartographie der Gegenwart in der Bundesrepublik Deutschland '84. Deutsche Gesellschaft für Kartographie, Bielefeld, BRD.

Reents E., 1957: Die neuzeitliche Luftfahrkarten und ihre Anwendungsbereiche. Karl-Heinz Meine – Eisenschmidt, Frankfurt, BRD.

Teichmann F., 2007: Network Focused Integration of Geospatial Data in Avionics. Masters Thesis, University of Krems, Krems, Austria, 56p.

Wilhelmy H., 1981: Kartographie in Stichworten. Ferdinand Hirt Verlag, Kiel, BRD.

Periodicals:

Bartels H., 2006: Digitale Kartengeräte für Eurofighter und Tornado. In Europäische Sicherheit 1/06 S7, E.S. Mittler u. Sohn, Herford, BRD.

Gassmann A., 1989: Verwendung digitaler Geländedaten in den Streitkräften. In Soldat und Technik 12/89 S873-880, Umschau Verlag Breidenstein, Frankfurt, BRD.

Gruender M., 2003: Cockpit der Zukunft – Hightech-Kommandozentrale. In Flugrevue Flugwelt 7/03 S102-105, Vereinigte Motorverlag, Stuttgart, BRD.

Hall T, 2006: Eclipsing the Past. In Flight International 5034/06 S60-67, Reed Business Publishing, Sutton, UK.

Hughes D., 2005: Civil avionics challenges. In Aviation Week and Space Technology 3/05 S195, McGraw Hill, New York, USA.

Hughes D., 2006: Civil Avionics on a rebound. In Aviation Week and Space Technology 3/06 S207-208, McGraw Hill, New York, USA.

Kant H. R., 1989: Verfügbare Milgeo-Datenbestände. In Wehrtechnik 7/89 S42-43, Wehr und Wissen Deutsche Gesellschaft Wehrtechnik, Bonn, BRD.

North D. M., 2002: F-35 Cockpit targets information integration. In Aviation Week and Space Technology 8/02 S52-53, McGraw Hill, New York, USA.

Penney S., 2000: Cognitive Cockpit – End of human pilot could be in sight. In Flight International 4748/00 S43-46, Reed Business Publishing, Sutton, UK.

Penney S., 2001: Network Fighter. In Flight International 4784/01 S118-124; Reed Business Publishing, Sutton, UK.

Phillips E. H., 2006: F-35 Cockpit Tech. In Aviation Week and Space Technology 6/06 S59, McGraw Hill, New York, USA.

Quaranta P., 2002: The evolution of visionics system. In Military Technology 7/02 S89-95, Moench Publishing, Bonn, BRD.

Schwarz K., 2004: Großes Bild für guten Durchblick. In Flugrevue Flugwelt 6/04 S78-83, Vereinigte Motorverlag, Stuttgart, BRD.

Scott W., 2000: Pilot-friendly cockpit benchmark of JSF design. In Aviation Week and Space Technology 3/00 S66-68, McGraw Hill, New York, USA.

Steve F. K., 2003: Cockpit. In AFM – Air Forces Monthly so12/03 S12-16, Key Publishing, Stamford, UK.

Thomalla V. K., 2006: Alles im Blick. In Flugrevue Flugwelt 2/06 S34-37, Vereinigte Motorverlag, Stuttgart, BRD.

Warwick G., 2001: Cockpit culture. In Flight International 4783/01 S39-40, Reed Business Publishing, Sutton, UK.

Internet:

URL 1 (2007):
International Civil Aviation Organization.
http://www.icao.int/ (11 may 2007).

URL 2 (2007):
Federal Aviation Administration.
http://www.faa.gov/ (11 may 2007).

URL 3 (2007):
Eurocontrol - European Organisation for the safety of Air navigation.
http://www.eurocontrol.int/index1.html (11 may 2007).

URL 4 (2007):
Jeppesen – Making every mission possible.
http://www.jeppesen.com/wlcs/index.jsp (11 may 2007).

URL 5 (2007):
Merriam-Webster Dictionary.
http://www.m-w.com/dictionary (11 may 2007).

URL 6 (2007):
UN Handbook on geographic information systems and digital mapping.
http://unstats.un.org/unsd/pubs/gesgrid.asp?mysearch=Handbook+on+geographic
+information+systems+and+digital+mapping (11 may 2007).

URL 7 (2005):
ICAO. International Civil Aviation.
http://www.icao.int (18 aug 2005).

URL 8 (2005):

Flight Services. Jeppesen Sanderson Inc.

http://www.jeppesen.com (18 aug 2005).

URL 9 (2005):

Aviation System Standards. National Aeronautical Charting Office (NACO).

http://www.naco.faa.gov (18 aug 2005).

URL 10 (2005):

Aviation Publication Service (APS). Del Mar.

http://www.apscharts.com (18 aug 2005).

URL 11 (2005):

Maps, Globes, Travel Guides. Map Town Ltd.

http://www.maptown.com (18 aug 2005).

URL 12 (2005):

Maps & Travel books from World of Maps. World of Maps.

http://www.worldofmaps.com (18 aug 2005).

URL 13 (2005):

Maptech. Maptech MapServer. Aeronautical charts.

http://mapserver.maptech.com (03 sept 2005).

URL 14 (2005):

National Geospatial-Intelligence Agency. Products and services.

http://www.nima.mil (03 sept 2005).

URL 15 (2005):

Friebe Flugbedarf. Luftfahrtkarten, manuals, pilot kits.

http://www.friebe-luftfahrtbedarf.de (03 sept 2005).

URL 16 (2005):

GCH Services, LCC – The Pilot Shoppe. Aeronautical Charts.

http://www.pilotshoppe.com (03 sept 2005).

URL 17 (2005):

Natural Resources Canada. Aeronautical and Technical Services.

http://aero.nrcan.gc.ca (03 sept 2005).

URL 18 (2005):

MapTrax Australia. Aeronautical Charts.

http://www.maptrax.com.au (03 sept 2005).

URL 19 (2006):

University of Texas Libraries. Maps.

http://www.lib.utexas.edu/ (22 april 2006).

URL 20 (2005):

Civil Aviation Authority. Aeronautical Charts.

http://www.caa.co.uk (03 sept 2005).

URL 21 (2005):

Federal Maps Inc. Aeronautical Charts.

http://www.fedmaps.com (03 sept 2005).

URL 22 (2005):

Civil aviation administration. Aeronautical Information Services.

http://www.ilmailulaitos.fi (03 sept 2005).

URL 23 (2005):

Federal Publications Inc. Maps and Charts. Aeronautical Charts.

http://www.fedpubs.com (03 sept 2005).

URL 24 (2005):

Map Connection Ltd. Aeronautical.

http://www.mapconnection.com (03 sept 2005).

URL 25 (2005):

Pilotfriend. Pilot resources directory.

http://www.pilotfriend.com (03 sept 2005).

URL 26 (2005):

Mypilotstore. Jeppesen and NACO Charts.

http://www.mypilotstore.com (03 sept 2005).

URL 27 (2005):

Airplane flight equipment. Charts.

http://www.afeonline.com (03 sept 2005).

URL 28 (2005):

Softchart. Aeronautical Products.

http://www.softcharts.com (03 sept 2005).

URL 29 (2005):

Marv Golden Pilot Supplies. Charts.

http://www.marvgolden.com (03 sept 2005).

URL 30 (2006):

International Virtual Aviation Organisation. FIR Charts & Procedures.

http://www.ivao.be/ (22 april 2006).

URL 31 (2006):

Eurocontrol. Cartography.

http://www.eurocontrol.int/index1.html (22 april 2006).

URL 32 (2006):

Luftfartsverket. Air Traffic.

http://www.lfv.se/ (22 april 2006).

URL 33 (2006):

Service de l'Information Aeronautique. Aeronautical Information.

http://www.sia.aviation-civile.gouv.fr/ (22 april 2006).

URL 34 (2006):

Austrocontrol. Air Traffic management.

http://www.austrocontrol.at/ (22 april 2006).

URL 35 (2007):

Deutsche Gesellschaft für Luft- und Raumfahrt.

www.dglr.de (11 may 2007).

Biography

Nationality:	Austria
Date of birth:	17. August 1964
Place of birth:	Salzburg, Austria

Occupational field:	Communication and Information Systems - CIS, Management and C4I Info Systems; Geographic Information Systems - GIS, GeoSciences, Global Environment.

Employment:

2006 to current:	Communication and Information Systems (CIS) section chief, MOD Austria, Dept. of Mng BH2010.
2002 – 2006	Communication and Information Systems Department Head (A6), Austrian Air Force HQ.
2002 to current:	Lecturer for Geoinformatics, University of Applied Sciences, Wiener Neustadt, Austria.
2001 to 2003:	External consultant for Geo- and GIS-implementation for disaster relief, Gisquadrat, Vienna.
2000 to 2002:	Chief geographic officer of the Austrian Air Force.
1996 to 2000:	Chief geographic officer of the Second Austrian Army Corps HQ.
1991 to 1996:	Geochemistry research Lab Manager and Post-Doctorate, University of Rochester, USA.

Education and training

2005 to 2007:	Telematik Management, Danube University Krems, Austria.

1999 to 2001:	Master in Advanced Studies; Geographic Information Systems; University of Salzburg, Austria.
1991 – 1995:	PhD and MSc; Geological and Environmental Sciences, Geochemistry; University of Rochester, USA.
1986 – 1988:	Master of Science; Petrology and Earth Sciences; University of Maine, USA.
1983 to 1989:	Magister rer. nat; Geological Sciences, Technical Geology; University of Salzburg, Austria.

Appendix

Due to University requirements, the German translation of the abstract is provided.

Netzwerkorientierte Integration von geographischen Daten in der Aeronautik

Zur effizienten und effektiven Durchführung von Luftraumoperationen müssen unterschiedlichste geographische Daten mit vordefiniertem Status bereitgestellt werden. Die standardisierten aeronautischen Flugnavigationskarten, An- und Abflugkarten und Flugplatzkarten werden durch georeferenzierte tabellarische Daten wie Standortidentifizierung (ID), Luftraumbegrenzungen, Luftfahrthindernisse oder Navigationspunkte ergänzt. Die resultierende luftfahrtspezifische Geodatenbank wird darüber hinaus durch einsatzrelevante geographische Daten wie Wegpunkte, Fernerkundungsdaten oder anlassbezogene Ergebnisse aus ortsbezogenen Services ergänzt. Die optimale Integration und Aktualisierung von Geodaten für die eingesetzten aeronautischen Informationssysteme sowohl am Boden als auch in der Luft sind eine wesentliche Voraussetzung für eine optimale Durchführung von Luftraumoperationen. Auf Grund der speziellen Erfordernisse bei Luftraumoperationen wie die verschiedensten Luftraumbenutzer, die dritte Dimension (z), die großen Distanzen, der notwendige Grad der Aktualisierung und Zertifizierung der Daten sowie der hohe Systemintegritätsgrad müssen die erforderlichen Geodaten besonders ausgearbeitet werden.

Die Entwicklung in der Bereitstellung von luftfahrtspezifischen Geodaten in der Aeronautik führte von einfachen analogen Navigationskarten über unabhängige analoge Geoinformationssysteme zu den gegenwärtigen digitalen Systemen mit gekoppeltem GPS-Anschluss. Die gegenwärtigen Trends gehen weiter Richtung integrierten oder vernetzten Avionicsystemen mit fortschrittlicher GIS-Funktionalität. Die aktuellen Geoinformationssysteme mit ihren modernen GIS-

Applikationen stehen in direkter Wechselwirkung zu den speziellen Erfordernissen der eingesetzten Geräteausstattung bzw. Hardware und sollten im Cockpit vollständig vernetzt sein. Diese Vernetzung würde den Datenaustausch der Avionic-Systeme z. B. für Navigationsdaten, genaue Zeitangabe und Flughöhe, aber auch, über eine zusätzliche Verknüpfung zum Flight Management System, den einsatzrelevanten Datenaustausch und den umfassenden Datenbankzugriff ermöglichen.

Die bedeutendste Innovationsverbesserung bei der Bereitstellung der Geodaten würde durch den Einsatz von netzwerkorientierten Verfahren eintreten. Basierend auf dem benötigten Datenfluss und Aktualisierungszyklus wären drei primäre Richtungen für eine Vernetzung von integrierten Geoinformationssystemen zu realisieren. Sichere Kommunikation, inklusive Datenübertragung, während des Fluges zur einsatzführenden Bodenstation würde service-orientierte Dienste mit anlassbezogenem Informationsmanagement für die Flugzeugbesatzung ermöglichen. Ein sicheres LAN oder Funknetz auf der Operationsbasis bzw. am Flugplatz könnte einen automatisierten Datentransfer zu und vom Flugzeug gestatten und dadurch sicherstellen, dass immer der aktuellste Datenbestand von Geoinformationen im Luftfahrzeug geladen ist. Offene Kommunikation mit vordefinierten Positionsangaben zwischen den Luftfahrzeugen oder zwischen Luftfahrzeug und Bodenstation würde Prozesse und Applikationen ermöglichen, die das Lage- und Situationsverständnis aller relevanten Einheiten massiv verbessern würde.

Um die Luftraumnutzung zu optimieren und gleichzeitig die Gefährdung zu reduzieren, muss Energie und Initiative mit Priorität aufgebracht werden, um die notwendigen luftfahrtspezifischen geographischen Daten produzieren und bereitstellen zu können. Diese Daten sollten in einem vernetzten Geoinformationssystem in die Avionic integriert werden und in netzwerkorientierte und auf Geodaten fokussierte Informationsprozesse optimal eingesetzt werden.

Printed by Books on Demand GmbH, Norderstedt / Germany